ԹՎԵՐԻ ՊԱՏՄՈՒԹՅՈՒՆԸ

THE NUMBER STORY

SMALL BOOK ONE

ENGLISH - WESTERN ARMENIAN

Numbers Teach Children Their Number Names

written and illustrated by

MISS ANNA

Early Reader Edition of *The Number Story 1*
Bronze Medal Winner, 2016 Wishing Shelf Book Award

Cover by | Lumpy Publishing
Layout by | Lumpy Publishing
Translated by Չեփիւր Չրքաբլեան
Coloring by Jieeun Woo and Maria Mirabella

Library of Congress Control Number: 2018902040

Names: Miss Anna, author.
Title: Number story : numbers teach children their number names / Miss Anna.
Description: Portland, OR: Lumpy Publishing, 2018.
Identifiers: ISBN 978-1-945977-95-4| LCCN 2018902040
Summary: The pictures and rhymes present stories which introduce numbers 0-10.
Subjects: LCSH Numeration—English—Western Armenian--Pictorial works--Juvenile literature. | BISAC JUVENILE NONFICTION /
Languages: English—Western Armenian
Classification: LCC QA141.3 .M57 2018 | DDC 513—dc23

Publisher: Lumpy Publishing
Website: www.missannabooks.com
Email: missanna@missannabooks.com

Paperback: ISBN 978-1-945977-95-4
Printed in the U.S.A.   1 3 5 7 9 10 8 6 4 2

Կ ուզե՞ս սորվիլ թիւերուն անունները:

It is very easy and a lot of fun!

Շատ դիւրին է և հաճելի:

Say-along our little jingle

Մեր հետ երգէ՛ մեր փոքրիկ երգը:

starting from Number One!

 Սկսմ́ նք Թիւ Մէկոɥ։

1

ONE looks like my one finger.

Մէկ

Մէկը նման է իմ մէկ մատին:

ONE!
ひとつ！

2

TWO trails a tail.

ԵՐԿՈՒ

Երկուն ունի պոչիկ:

A TAIL! ጣበ2Ꮆ Ꮞ:

3

THREE has bumps.

ԵՐԵՔ

Երեքը կոր է:

Նայի՛ր անոր կոթերուն:

4

FOUR   carries a sail.

ՉՈՐՍ

Չորսը առագաստանա՜ւ է:

A SAIL!
ԱՌԱԳԱՍՏ։
ԱՌԱԳԱՍՏԱՆՆԱՒ։

5
FIVE   is a racing track.
ՀԻՆԳ
Հինգը մրցուլի ճանապարն է:

VROOM

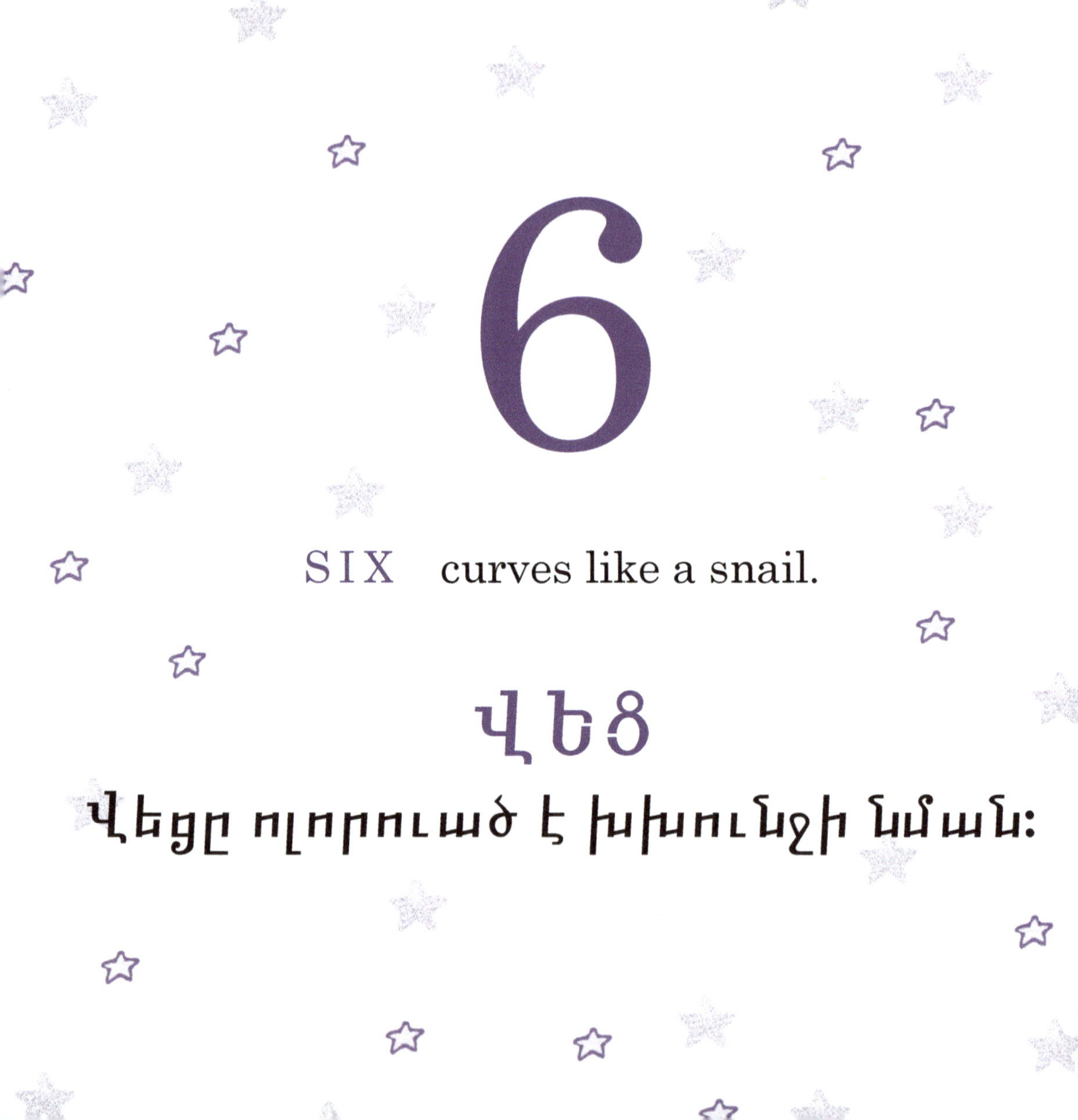

6

S I X   curves like a snail.

ՎԵՑ

Վեցը ոլորուած է խխունջի նման:

A SNAIL!   Խխունձ:

7

SEVEN has a sharp angle.

Եօթը

Եօթն կացին է:

BE CAREFUL! IT'S SHARP!

ՈՒՇԱԴԻՐ Ր: ՍՈՒՐ Է:

8

EIGHT   is rollercoaster rails.

ՈՒԹ

Ութն ուրախ երկաթուղի է:

 uhrruuuuu:
YIPPEE!

9

NINE   is a bubble on a stick.

ԻՆՆ

Իննը պղպջակ է փայտի վրայ:

A BUBBLE! ՊղՊձԱՙ Կ:

10

TEN is an eye of a whale.

ՏԱՍԸ

Տասըն կետի մէկ աչքն է։

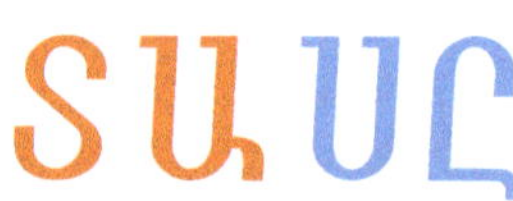

HELLO!

ԲԱՐԵՒ:

And

Կ

O

ZERO   is an empty pail.

ԶԵՐՕ

Զերօն դատարկ դոյլ է:

IT'S EMPTY!
ԴԱՏԱՐԿ Է:

Thank you for playing with us today.

We had a lot of fun too!

Շնորհակալ ենք, որ այսօր մեր հետ խաղացա՛ք:

Շա՛տ ուրախացանք:

We are your Number friends,
Zero to Ten,
Who will be here for you~

Մենք ընկերներդ ենք

Զերոէն Տասը։

Մենք միշտ հոս ենք, հետդ ենք։

Bye-bye now!
See you again soon!

Յաջողութի՛ւն։

Շուտո՛վ կը հանդիպինք։

The Numbers are *SINGING* too!

To sing-a-long, look for Miss Anna Number Story
at your favorite music store like iTUNES.

MP3

| Numbers 0-10 IDENTIFYING & COUNTING | Numbers 11-20 & Ordinals first, second, third... | Numbers 0-100 & Place Values ones, tens, hundreds... | About Clocks & Telling Time hours, minutes, seconds |

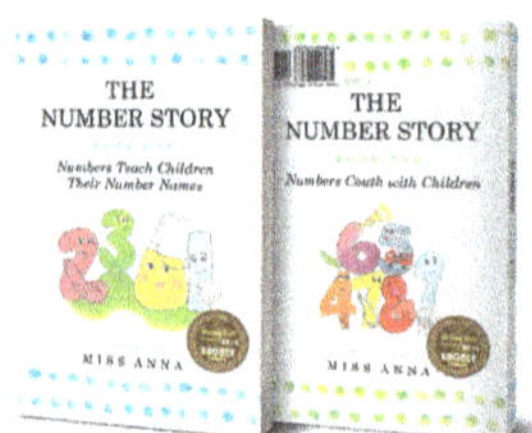   

| Number Story 1 & 2 | Number Story 3 & 4 | Number Story 5 & 6 | Number Story 7 & 8 |
| isbn: 978-0-996216-48-7 | isbn: 978-1-945977-01-5 | isbn: 978-1-945977-06-0 | isbn: 978-1-949320-40-4 |

For more Miss Anna books to love,
visit us at

www.missannabooks.com

Numbers are working hard all over the world!
Come Travel the World with Us!